ISBN 978-1-331-74746-8
PIBN 10229513

1 MONTH OF
FREE
READING

at

www.ForgottenBooks.com

By purchasing this book you are eligible for one month membership to ForgottenBooks.com, giving you unlimited access to our entire collection of over 700,000 titles via our web site and mobile apps.

To claim your free month visit:
www.forgottenbooks.com/free229513

English
Français
Deutsche
Italiano
Español
Português

www.forgottenbooks.com

Mythology Photography **Fiction**
Fishing Christianity **Art** Cooking
Essays Buddhism Freemasonry
Medicine **Biology** Music **Ancient
Egypt** Evolution Carpentry Physics
Dance Geology **Mathematics** Fitness
Shakespeare **Folklore** Yoga Marketing
Confidence Immortality Biographies
Poetry **Psychology** Witchcraft
Electronics Chemistry History **Law**
Accounting **Philosophy** Anthropology
Alchemy Drama Quantum Mechanics
Atheism Sexual Health **Ancient History**
Entrepreneurship Languages Sport
Paleontology Needlework Islam
Metaphysics Investment Archaeology
Parenting Statistics Criminology
Motivational

THE FOURTH DIMENSION
AND THE BIBLE

BY

WILLIAM ANTHONY GRANVILLE, Ph.D., LL.D.,

PRESIDENT OF GETTYSBURG COLLEGE.

Formerly Instructor of Mathematics in Yale University; Author of Granville's Differential and Integral Calculus, Granville's Plane Trigonometry, Granville's Spherical Trigonometry, and Granville's Logarithmic Tables; Joint Author of Smith and Granville's Elements of Analysis; Inventor of Polar Coordinate Plotting Paper, Yale System of Mathematical Note Books, and Granville's Transparent Combined Ruler and Protractor.

ARTI et VERITATI

BOSTON
RICHARD G. BADGER
THE GORHAM PRESS

Made in the United States of America

The Gorham Press, Boston, U. S. A.

PREFACE

That philosophy and the physical sciences, when called to the defence of Christianity, have so often proven to be broken reeds has been a keen disappointment to many sincere seekers after the truth. A little reflection, however, will make clear that in the very nature of things such failures must inevitably result, because structures resting on shifting sand can never serve as strong buttresses. No two systems of philosophy agree and every philosopher is continually tinkering with his own system. There is no final authority in philosophy to which an appeal can be made. Such being the case it is obviously not to be expected that philosophy can furnish us with proofs of the truths of Christianity whose validity will be generally recognized. And the sciences have also fallen short for similar reasons. The foundations of our physical sciences are far from stable. To doubt the principles of the conservation of energy, the conservation of matter, or the indivisibility of the atom, would have been considered as rank scientific heresy a few years ago; today all physicists are familiar with phenomena which contradict one or more of them. The Einstein Theory of Relativity is now opening up new fields for scientific investigation which promises results that will practically revolutionize many current fundamental scientific concepts. In fact, the foundation principles on which the physical sciences rest have been changing so rapidly of late that practically all the science text-books now used in our schools are more or less obsolete. The physical sciences as well as philosophy are in a constant state of flux and flow as well as of growth, and this must necessarily continue to be the situation until the ultimate truth is reached, if it is ever attained, in these domains of knowledge. As all will readily agree that the limits of these regions have not yet been reached and, what is more, that these limits are not even measurably in sight, it is evident that we cannot reasonably expect that much constructive light will be thrown on Christianity now or in the near future by either philosophy or the physical sciences.

To satisfy completely human reason any confirmation of our Christian beliefs must come from a source whose authority no one can question. Is there then any department of human knowledge whose foundations rest, not on shifting sand, but on the bed rock of absolute truth? Pure mathematics satisfies this condition; it is the only exact science that God has revealed to man and the truths which it contains are the only truths that can be absolutely established thru pure reason. Because pure mathematics reveals absolute truth it is part of God Himself, for God is the essence of all truth. It was Emerson who said, "nature geometrizes," and as nature is the handiwork of God we may say that God geometrizes. In view of this it is rather surprising that so few attempts should have been made in the past to throw light from mathematics on the fundamentals of Christianity. This is practically a virgin field for theological research.

Pure mathematics is the vanguard of all the sciences, it has marched centuries in advance of them all. For instance, Calculus was discovered nearly two hundred years before the science of Chemistry was far enough advanced to allow the use of Calculus as one of its instruments of research. Practically the same thing is true of all the other physical sciences, including engineering. It may be, therefore, that the science of theology is not yet far enough advanced to admit of the ready use of mathematics as an instrument of research. Today mathematicians are forging ahead into realms of abstract thought and pure logic as never before; judging from the past, some of the truths which they are revealing now may not be put to what we are pleased to call practical use for centuries to come.

That mathematics will ultimately prove to be a valuable aid in solving many of the perplexing problems connected with our Christian religion is the firm belief of many sound thinkers. Professor C. J. Keyser has pointed out many striking correspondences between the transcendental properties of some elementary mathematical concepts and those transcendental attributes of God and His creation which are intimately associated with our religious beliefs. Concerning the possibility and the probability of future investigations along this and other lines shedding new light from mathematical sources on current religious thought he has the following to say in *Hibbert Journal*, Vol. VII: "Over and above the humbler rôle of mathematics as a metrical and computatory art, over and above her unrivalled value as a standard of exactitude and as an instrument in every field of experimental and observational research, even beyond her justly famed disciplinary and emancipating power, releasing the faculties from the fickle dominion of sense by winning their allegiance to the things of the spirit, inuring them to the austerities of reason, the stern demands of rigorous thought, giving the mental enlargement, the peaceful perspective, the poise and the elevation that come at length from continued contemplation of the universe under the aspects of the infinite and the eternal—my conviction that above and beyond these services, which by common consent of the competent are peculiarly her own, mathematics will yet further demonstrate her human significance by the shedding of light more and more copious as the years go by on ultimate problems of philosophy and theology, is not a passing fancy, or a momentary whim."

The only motive of the author in writing this book is his sincere desire to throw some light, even though very dim, on some of the questions connected with our Christian beliefs. This is not, however, an endeavor to construct a system of Christian theology on mathematical foundations. The author's chief aim will be to point out the remarkable agreement which exists between numerous Bible passages and some of the concepts which follow quite naturally from the mathematical hypothesis of higher spaces. In making this attempt the author is well aware that he is on a "no man's land" exposed to fire from the mathematical trenches on the one side and the theological trenches on the other. But "fools step in where angels fear to tread" and sometimes they survive to tell the tale.

The scripture passages quoted in this text are not employed for exegetical purposes but to show that the conclusions arrived at in no way contradict the divine word.

While there is quite an extensive bibliography

dealing with the subject of higher spaces, the liter-
ature bearing on their relation to Bible statements is practically nil. After a thoro search only
scattered references, each embracing only a few
phrases or sentences, have been found which
might be construed as supporting some of the
conclusions arrived at in this text. The author
is therefore taking a voyage on practically un-
charted seas. No doubt it may seem presumptuous on the part of the author, who aspires to
be a mathematician and is no theologian at all,
to poach thus on theological preserves; but the
case is perhaps no worse than if a theologian who
is not also a mathematician should undertake the
task.

A sincere effort has been made to present the
matter in popular form. To understand the text it
is not necessary for the reader to be either a pro-
fessional mathematician or a profound theologian.
No extended excursion into the theory of higher
spaces will be attempted; the author has limited
himself to those fundamental concepts only which
are necessary for the illumination of the applica-
tions to be made. No proofs will be given of the
mathematical propositions hereinafter stated; but
the reader can rely on them absolutely, no matter
how startling the conclusions may appear. They
have all been proven by mathematicians of the
highest repute and the validity of these proofs

are not more in dispute than is the proposition that $2 \times 2 = 4$.

The use of the term "Fourth Dimension" as part of the title of this book does not mean that we shall confine our discussion to that space exclusively, important as it is; it is so used because in the popular mind this term represents the mathematical concept of higher spaces in general.

WILLIAM ANTHONY GRANVILLE.

Gettysburg College,

Gettysburg, Pa.

CONTENTS

THE FOURTH DIMENSION
AND THE BIBLE

THE FOURTH DIMENSION AND THE BIBLE

CHAPTER I

THE CONCEPT OF SPACE

(a) *Zero-Dimensional Spaces.* A·

A point, as A, may be conceived of as a space of no dimensions. Every point in that space coincides with A and therefore is at no distance from it. A point cannot move about in that space, hence its degree of freedom of motion is zero. Since we know that there is an infinite number of points there is also an infinite number of such zero-dimensional spaces.

(b) *One-Dimensional Spaces.*

Let O be a fixed point on a straight line. To determine the position of any point on that line, as A, all we need to know is its distance to the right or left of O expressed in terms of some unit of length, a positive number to indicate that it is so far, say, to the right of O, or a negative number when it is to the left of O. A single number will therefore suffice to determine the position of any point on the straight line; hence the straight line is said to be one-dimensional. ,The straight line may also be conceived of as a space of one dimension; any point in it has only one degree of freedom of motion, it can move only back and forth in the line.

The circumference of a circle may be considered as a curved one-dimensional space, as well as every other curved line, whether it returns into itself or not, because (as in the case of the straight line) the position of any point, as A, may be determined by a single number representing its distance measured along the curve from a fixed point (as O·) on that curve. Obviously there is an infinite number of one-dimensional spaces because the number of straight and curved lines is infinite.

(c) *Two-Dimensional Spaces.*

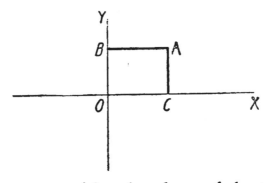

Let us now consider the plane of the two mutually perpendicular lines OX and OY intersecting at O. Any point in this plane, as A, may be determined by its perpendicular distances BA and CA from OY and OX. Calling distances from OY and OX measured to the right and upward as positive, and distances to the left and downward as negative, the position of any point in the plane will be determined by the two numbers representing these distances; hence the plane is said to be two-dimensional. This plane may also be conceived of as a space of two dimensions; any point in it has two degrees of freedom of motion, it can move to the right or left, and up or down. A point in the plane can move from one position to any other in that plane by moving in a path which is always parallel to one of two fixed mutually perpendicular lines in that plane. Take, for instance, the top of a rectangular table. Any article on it, as an ink stand, may be moved from one spot to any other along a path that will always be parallel to one of the edges of the table top.

The surface of a sphere may be conceived of as a curved two-dimensional space. The position of any point on the surface of the earth is determined when we know the two numbers indicating its geographical latitude and longitude. By moving along parallels of latitude and lines of longitude we may travel from any one place to any other. Not only the surfaces of all spheres but also an unlimited number of other curved surfaces may be conceived of as curved two-dimensional spaces. Evidently there is an infinite number of two-dimensional spaces.

(d) *Three-Dimensional Spaces.*

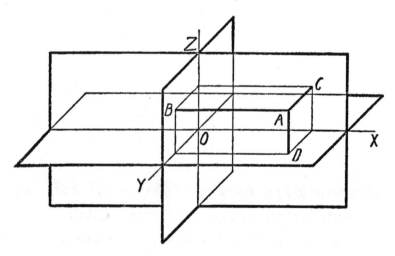

The space in which we move and have our being is three-dimensional; we say that bodies in it have length, breadth and thickness. Consider three mutually perpendicular planes meeting at O. The position of any point in our space, as A, is then determined by the three numbers representing its perpendicular distances BA, CA, DA, with the proper signs, from the three planes. A point in our space has three degrees of freedom, it can move to the right or left, forward or backward, and up or down. For instance, an article in a rectangular shaped room may be moved from one place to any other by moving it parallel to the directions of the length, breadth, and height of the room. The reader can reach no place which may not be reached by going north or south, east or west, and upward or downward.

The space we live in is the only three-dimensional space of which we have any knowledge gained by experience. We are not warranted, however, in making the statement that no other three-dimensional spaces exist; such a proposition cannot be proven. Curved lines have given us the conception of curved one-dimensional spaces, and curved surfaces the conception of curved two-dimensional spaces. We may also conceive of the existence of curved three-dimensional spaces although in this case we have no corresponding

visual images on which to lean; we cannot construct in our minds the picture of a curved three-
dimensional space.　From analogy it is evident
that there may be an infinite number of three-
dimensional spaces, both curved and not curved.

(e) *Spaces of Order Higher than the Third,*
also called Hyper-Spaces.

In like manner we define a four-dimensional
space to be one in which it requires four numbers
to determine the position of any point.　A point
in such a space will have four degrees of freedom.

A five-dimensional space would be one in which
it requires five numbers to determine the position
of a point; a point in it would have five degrees
of freedom.　Similarly we define spaces of 6, 7,
8, . . . n, dimensions.

It is impossible for us to form a mental picture
of, say, the fourth dimension, the idea is wholly
intangible.　Nevertheless, it is not meaningless
or an absurdity, but a useful mathematical concept which has led to the development of a geometry of four dimensions involving no contradic-
tions.　We reach this new concept of higher
spaces not at a single leap but slowly and gradually by climbing a ladder whose lowest rungs are
the lower dimensional spaces with which we are

in a certain sense familiar. Guided by analogy with dimensions of a lower order we shall thus try to arrive step by step to at least a partial and symbolic idea of the meaning of higher dimensions. Of these we will be chiefly concerned with the fourth dimension. Some claim that the Bible recognizes space of four dimensions in Job 11:8, 9 and Eph. 3:18 where length, breadth, depth and height, i. e., four dimensions, are mentioned.

CHAPTER II

GEOMETRIC UNITS IN EACH SPACE

That there exist intimate relations closely approaching a sort of interdependence between the spaces of various orders is forcibly suggested by the following discussion showing how the unit in each space may be generated by the unit of the space of the next lower order moving in a new direction not contained within itself.

Starting with zero-dimensional space (a point) we may consider the point itself as the unit.

A   POINT

**Zero-dimensional
Space Unit**

If the point A moves the distance of, say, one inch to the right it will generate the unit line AB.

LINE

A━━━━━━━━B   Bounded by the points A and B
One-dimensional Space   (zero-dimensional units)
Unit

If the line AB moves upward one inch in a

direction perpendicular to itself it will generate the unit square ABCD.

SQUARE

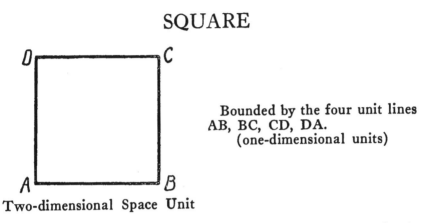

Bounded by the four unit lines
AB, BC, CD, DA.
(one-dimensional units)

Two-dimensional Space Unit

If the square ABCD moves backward one inch in a direction perpendicular to itself it will generate the unit cube ABCDEFGH.

CUBE

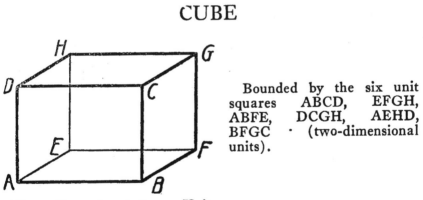

Bounded by the six unit squares   ABCD,   EFGH, ABFE,   DCGH,   AEHD, BFGC · (two-dimensional units).

Three-dimensional Space Unit

If the cube ABCDEFGH moves away one inch in a direction perpendicular to itself it will generate the corresponding unit in four-dimensional space. This unit has been given the name *tesseract*. Here we meet our first serious difficulty. We are familiar with the idea of motion perpendicular to a line, and of motion perpendicular to a square (or, what amounts to the same thing, of motion perpendicular to two mutually perpendicular lines lying in the plane of the square), but we cannot conceive of a motion perpendicular to a cube (i. e., a single motion perpendicular to all the faces of the cube, or, as it may also be stated, perpendicular to any three adjacent edges of the cube). Not one of our senses calls for a fourth direction perpendicular to the other three, experience leaves us satisfied with three dimensions. We cannot visualize the tesseract but it is possible for us to construct a figure in our space which will symbolize the tesseract, that is, a figure having many of the properties of the tesseract. Such a figure is the following which we may conceive of as having been generated by the original cube ABCDEFGH expanding at a uniform rate into the cube A′B′C′D′E′F′G′H′, the directions AA′, BB′, CC′, DD′, EE′, FF′, GG′, HH′, being supposed perpendicular to our space and equal in length to an edge of the cube.

In generating a tesseract by the motion of a cube, the latter's corners generate edges, its edges generate faces (squares) and its faces generate

TESSERACT
(Symbolical)

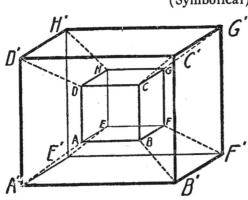

Bounded by the eight unit cubes ABCDEFGH, A'B' C'D'E'F'G'H', A'B'BAE' F'FE, DCC'D'HGG'H', BB'C'CFF'G'G, A'ADD' E'EHH' A'B'C'D'ABCD, EFGHE'F'G'H' (three-dimensional units).

Four-dimensional Space Unit

cubes. The faces (squares) generated, ABB'A', BFF'B',DCC'D', etc., and the cubes generated, A'B'BAE'F'FE, BB'C'CFF'G'G, etc., necessarily appear distorted in the above figure. The number of the various elements of the tesseract are indicated in the following table which may be easily verified by the reader.

	Number in Initial Cube.		Number Generated.		Number in Final Cube.		Number in Tesseract.
Corners	8	+	0	+	8	=	16
Edges	12	+	8	+	12	=	32
Faces	6	+	12	+	6	=	24
Cubes	1	+	6	+	1	=	8

The statements made opposite the figures on pp. 18, 19, 20 indicate that each space unit is bounded by space units of the next lower order.

The following table indicates the number of each of the various elements that are associated with the boundaries of the various space units.

	Points.	Lines.	Squares.	Cubes.
Zero-dimensional unit (point) ..	1	0	0	0
One-dimensional unit (line)	2	1	0	0
Two-dimensional unit (square) .	4	4	1	0
Three-dimensional unit (cube) .	8	12	6	1
Four-dimensional unit (tesseract)	16	32	24	8

Since there is an infinite number of points in a line, an infinite number of lines in a square, and an infinite number of squares in a cube, it is near at hand to assume that a space of any order contains an infinite number of lower dimensional spaces. The following definition of space has therefore been suggested:

Space is that which separates two portions of the next higher space from each other.

Let us illustrate what is meant.

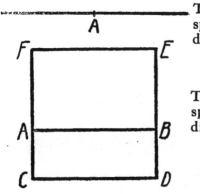

The point A (zero-dimensional space) divides the line (one-dimensional space) into two parts.

The line AB (one-dimensional space) divides the plane CE (two-dimensional space) into two parts.

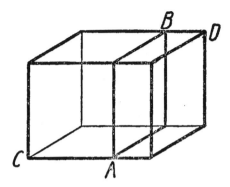

The plane AB (two-dimensional space) divides the solid CD (three-dimensional space) into two parts.

Analogy now steps in and suggests that we carry this one step further by making the statement that *our three-dimensional space probably divides four-dimensional space into two parts.* And so on for the other higher-dimensional spaces.

CHAPTER III

ON THE EXISTENCE OF HIGHER SPACES

It is to be hoped that by this time there has dawned in the mind of the reader at least a faint idea of the concept of higher spaces, an inkling of what is meant by, say, the fourth dimension. Additional discussions from other viewpoints tending to show the probability of the existence of space of four dimensions will be given in later chapters of this book.

We experience no difficulty in forming a geometric conception of zero-dimensional space (point), of one-dimensional space (line), of two-dimensional space (plane), and of three-dimensional space (our universe) ; but when at first we try to give a geometric interpretation to space of four, five, six and higher dimensions, we are at sea. We find nothing tangible on which to lean in our fund of experiences accumulated in dealing with the material things in our world, things of which we have become conscious thru our senses. We are therefore tempted to at once arrive at the conclusion that no spaces of four, five, six, etc., dimensions exist. Such a conclusion, however, would be premature because, as has already been stated, the proposition that higher spaces do not exist cannot be proven.

Firstly, to dismiss the concept of the existence of higher-dimensional spaces with the statement that such a thing is incomprehensible is begging the question, unless we are also willing to abandon the idea of the existence of our own space of three dimensions, because its properties are also in the main unfathomable mysteries to us. For instance, "it may properly be claimed that a three-dimensional solid of infinite length, infinite breadth, and infinite thickness would embrace infinite space; but is it possible to picture or comprehend what infinite space is? Can a finite mind picture a space with no beginning and no ending; limitless space in which our vast solar system is a mere dot, in which the known stellar universe is probably also comparatively a mere dot, altho it is actually so vast in extent that the light from some of its component stars which started towards us generations ago or centuries ago is only now reaching us? If our space is limitless, the idea is incomprehensible, and if it is limited, its limits are incomprehensible. Our space is limited or it is limitless; in either case the idea is incomprehensible. Thus the mere statement that a concept is incomprehensible does not prove its non-existence." (Gunnell.) Obviously, it would be absurd for us to doubt the existence of our own space simply because many of its properties are incomprehensible to us. The human mind cannot fully encompass the idea of infinity and yet this is one of the most useful mathematical concepts employed daily without hesitation in the teaching of elementary as well as advanced mathematics.

Secondly, only an ignoramus would claim that nothing exists which does not lie within his realm of understanding, knowledge and experience. One hundred years ago the results now achieved thru the telephone, the phonograph, or the wireless telegraph and telephone would have been inconceivable, would have appeared as miracles even to the wisest men of that day.

Thirdly, probably few readers of this book would hesitate to make the assertion that one-dimensional spaces (lines) and two-dimensional spaces (planes) do actually exist. But as a matter of fact, the line as a one-dimensional space and the plane as a two-dimensional space has no more of a real existence to us than has four-dimensional space. No one has ever seen or heard or felt or tasted or smelled either a line or a plane. All our sense perceptions and the experiences due to them are three-dimensional because all material things are three-dimensional. Our senses can perceive only material things and those only thru our sensory nerves which are also material and therefore three-dimensional. Even the shadow on a wall cannot truly be said to be two-dimensional because what we really observe is that fewer rays of light enter our eyes from the region covered by the shadow, while from the region surrounding the shadow many rays of light do enter our eyes. And these rays of light are reflected from material particles (pigment) on the surface of the wall, that is, from particles of matter (three-dimensional). Our eyes communicate to our brains only three-dimensional knowledge; *"for the comprehension of anything which has only one or two dimensions a purely intellectual act of abstraction must be added to that of perception."* Lines drawn on paper, strings, rods and other objects which suggest to us the geometric idea of a line are really three-dimensional forms of matter in which two dimensions are very small as compared with the third, its length. Similarly the plane surfaces of our experience (sheets of paper, blackboard, etc.) are all three-dimensional forms in which the thickness is very small as compared with the length and breadth; but these helps have been found useful to us in visualizing the geometric idea of a plane. Thus lines drawn on paper or on a blackboard have been found very helpful in teaching plane geometry, *but they are not a part of plane geometry itself*. If all such material aids were discarded, or should cease to exist, plane geometry as we know it would still remain, and every theorem proven in the plane geometry textbooks used in our schools today would still hold true. *Four-dimensional geometry exists in the same sense that plane geometry exists,* both have a real existence. The difficulty we encounter in our attempts to form mental images of geometric figures of four dimensions consists in the fact that we have no material aids to our imagination, while figures in plane and solid geometry are readily conceived of as corresponding to many material objects in the world with which we are familiar.

Geometry itself, therefore, throws no light on the real nature of any space, not even on the nature of the three-dimensional space in which we live. In fact, it has not even been proven that the solid geometry taught in our schools and the space in which we live are in exact correspondence; all we can say is that as far as our observations have extended, the one does appear to fit the other. The study of space is an empirical science, its conclusions must be arrived at thru observation and experiments, while geometry is a science built up by pure reason, it is a branch of pure mathematics. To the mathematician the geometry of four dimensions, as well as the geometry of two dimensions, has just as real an existence as the geometry of three dimensions, the geometry relating to the space of our experience. Many extended treatises have been written on the geometry of higher spaces by mathematicians of the highest repute and the validity of the results obtained by them are not questioned. *There is no doubt about the existence of geometry of four dimensions, but it has not been proven that there exists a corresponding space of four dimensions.* Therefore if such a thing as four-dimensional space does exist, we already have at hand a geometry to fit it. It is important that the reader should clearly perceive that the concept of space is one thing, while the geometry of space is quite another.

The geometry of four dimensions having then a real existence in the same sense that geometry of one, two, or three dimensions exists, the question naturally arises, does there exist a four-dimensional space which fits this geometry in the same sense that the three-dimensional space of our experience fits our geometry of three dimensions? This question is one which we cannot escape, it will continue to haunt the thoughtful student, it is the most interesting question connected with the subject "The possibility that we are a part of a four-dimensional space with physical limitations which confine us to a three-dimensional space, and with limitations of our senses which prevent us from perceiving anything outside of this space—this possibility excites the interest of all who are inclined to abstract speculation. Attempts may be made to discover physical proofs, of such a space, to build up theories on its basis that will explain discoveries of modern physics as yet but little understood, or by it to account for various mysterious phenomena. Most of us are satisfied that no real proofs of the existence of space of four dimensions will be found along these lines. Even a workable hypothesis based on the (assumed) existence of four-dimensional space, tho it might serve temporarily better than any other hypothesis, would hardly justify a belief in this existence. But we do say that the existence of space of four dimensions can never be disproved by showing that it is absurd or inconsistent; for such is not the case. Nor, on the other hand, will the most elaborate development of analogies of different kinds ever prove that it does exist." (Manning.)

Kant not only recognized the possibility of the existence of spaces of more than three dimensions, but he inferred their very probable real existence. "If it is possible," he says, "that there are developments of other dimensions of space, it is also very probable that God has somewhere produced them. For his works have all the grandeur and variety that can possibly be comprised." Or, as a mathematician might put it, why should God have created geometry of, say, four dimensions, if there exists nowhere a space that will fit it? God does nothing that is futile or unnecessary.

The following quotation from "Creation Ex Nihilo" (Gruber) will aid the reader in maintaining the proper attitude of mind towards what has gone before as well as prepare him for what shall follow

"Who can limit the possible existence, within and beyond the physical universe, even of *other* beings altogether inconceivable by us? Such transcendent entities, as also God and the human soul, because not governed by physical laws, could in no way affect our physical sensorium. Not being limited by time and (our) space relations, they might co-exist with, and around, us and thruout, as well as beyond, the physical universe—and this without in the least affecting human experience and consciousness. . . . As experience, thru its proper avenues of approach, and consciousness, have their necessary limitations, as sources of knowledge, in their limited points of contact with self and nature, so their bodyings forth in the pronouncements of physical science do not exhaust reality. Hence, the great whole of reality can never come within their compass. And this is even true of *physical* reality. But, as to the transcendent hyperphysical or spiritual realities, consciousness and experience, unaided—and therefore physical science—would forever have to remain without a point of contact, and therefore in total ignorance. No one should, therefore, attempt either to set a limit to existence by his limited experience or make his finite reason the measure of the immeasurably complex universe of the infinite and eternal God."

"A person born without the sense of sight cannot see light and color. He cannot even form any real *conception* of them. To him all is darkness. And, as far as he could by his own powers discover, both light and color would have no existence or would belong to the same category as darkness. The man born without the sense of hearing can neither hear sound nor can he even form any correct *conception* of sound and music. And thus, to a (congenital) blind and deaf individual, even this very tangible physical world is an entity altogether different from the reality. Many of its marvelous phenomena of beauty are to him totally non-existent. And if, perchance, some explanation or revelation to him of these things were attempted by a seeing and hearing personality, these phenomena would yet in a sense be utterly inconceivable by him. He can only by *touching,* etc., acquire some indefinite idea of *grosser* forms and movements. But he could not perceive even any *effect* of those subtle marvelous vibrations that produce light and color and music."

"Such an individual lives in a world of marvelous beauty, but he beholds it not, nor can he even form any proper conception of it. But, surely, it would be an almost unpardonable presumption on his part to deny the existence of the glorious rainbow in the heavens and to argue with an entranced auditor against the existence of the majestic symphonies of a Beethoven. For these things that lie beyond his limited physical senses, he must needs accept the testimony of those who have the necessary *senses of perception* to know their reality."

"Like the blind and deaf individual, we stand amid the wonders of nature. Tho we can perceive, in light and color and music, a minute fraction of the *effect* produced by waves of ether and waves of air, yet these physical *waves themselves* lie totally beyond even *our* natural sense-organs. Like that blind and deaf individual, who thru his sense of touch, etc., can form some idea of his physical environment, so we with our limited sense-organs can acquire some knowledge of the surrounding universe. But, as in his case, our limited senses permit of but a very *partial* knowledge, and beyond their range there are realities even in *physical nature* concerning which we can only speculate and which we may never know. We are like children watching a game, from a point some steps away, thru a small crack or knot hole in a boarded enclosure. All appears fragmentary and partial. Immeasurably the greater part of the universe, from the infinitesimally small to the universal whole, in ten thousand marvels, lies beyond the range of the whole outfit of our physical senses, or is a physically intangible reality. Thus, the visible light or color spectrum constitutes the record of but a minute fraction of the whole range of the mysteriously wonderful perpetual dance of the imponderable ether, it constitutes but one of the octaves of the many-octaved key-board of vibrations. And yet, the other octaves no less truly *exist* as with our own invented tools of investigation we are more and more discovering from their effects; but these *octaves themselves* we cannot perceive. If our eyes could be so adjusted as to enable us to behold the whole range of this fundamental reality, vistas of visions hitherto inconceivable would lie before our astonished gaze. . . . And so might we conceive of superadded transcendent *spiritual faculties,* if not confined within our physical organism. And thus the transcendent glories of spiritual realities, of angelic beings, and even of God, might entrance the astonished perceiving personality."

In what follows in this text, higher dimensional spaces, and in particular the fourth, will often be referred to as if they really existed. The reader should clearly understand that this is done only to save words by avoiding a needless repetition of the statement that their existence is only assumed.

CHAPTER IV

ABOUT THE SUPPOSED INHABITANTS OF OTHER SPACES THAN OUR OWN

As we are now going to make several excursions into the realm of pure thought and abstract speculation it will be necessary for the reader to approach the subject with an open and receptive mind. He will need to occasionally cut loose from the sense perception moorings which anchor him to our material world, he will often have to leave experience behind and reason wholly from analogy. A truly sympathetic attitude will add much to the benefits that may be derived from the discussions which follow.

Suppose that one, two, and four-dimensional spaces do exist, in the same sense that the three-dimensional space of our experience exists, and let us imagine that in each of these spaces beings exist, each individual being restricted to his own particular space in his movements, his perceptions, and his understanding, just as we are restricted to a three-dimensional space. Then let us by analogy try to conceive what the attitude of a four-dimensional being living in a four-dimensional space would be towards us living in space of three dimensions by considering what our attitude would be towards beings living in space of one or two dimensions just as if we were really conscious of their existence. For the sake of brevity we shall designate space of two dimensions as Flatland and its inhabitants as Flatlanders, and the space of one dimension as Lineland and its inhabitants as Linelanders. Corresponding to the space in which they live Linelanders would have one dimension only (length) and Flatlanders two dimensions only (length and breadth).

FLATLAND

Let us consider Flatland. By imagining two-dimensional beings (Flatlanders) living in a plane (Flatland) and able to glide about in that plane but absolutely unable to move out of it or to perceive anything of a third dimension, just as we are unable to perceive anything of a fourth dimension, we may obtain a vivid idea of our relation to a possible four-dimensional space. A consideration of what would be the attitude of Flatlanders towards any conception of space of three dimensions makes clearer what should be our attitude towards the conception of a space of four dimensions. In doing this we may suppose that what we call two-dimensional matter is really three-dimensional, but with a very slight thickness in the third dimension which Flatlanders are unable to recognize and of which they are entirely unconscious. Thus we may imagine Flatland in the nature of a soap film (formed in the closed loop of a wire) in which Flatlanders live. Or, if we are willing to accept the statement of some philosophers and physicists that a particle of our matter has no real substance (in the sense in which we ordinarily use that term) but consists only of a bundle of forces, attractive and repellant, then there is no difficulty in thinking of such forces as lying entirely in a plane (or in a line). All bodies in our space are bounded by surfaces, in Flatland all bodies have one-dimensional boundaries (lines and curves). To the skin (surface) of a human being there would correspond the bounding contour (perimeter) of a Flatlander. To other Flatlanders he would be exposed only along his perimeter, his interior could be reached by them only thru an opening in this perimeter. A Flatlander surgeon operating on a Flatlander would have to make an incision in the perimeter of his patient just as one of our surgeons must make an incision in the skin of a patient to remove his appendix. To one of our surgeons gifted with two-dimensional vision, however, all the interior organs of a Flatlander would be exposed and he could operate without cutting his Flatlander patient open. Not only could we see (if we had two-dimensional vision) the insides of Flatlanders but we could also see the insides of their houses and all their closed compartments. To illustrate let us assume that the plane ABCD on this page is part of Flatland and that a Flatlander inhabiting it is triangular shaped (Fig. a).

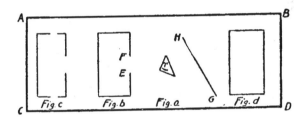

As there is no "upward" or "downward," no "above" or "below" in Flatland, houses and barns would have no roofs or floors, only the lines of their sides (perimeters) would be there. Figure (b) may be taken to represent the house of a Flatlander. He could enter his house only thru the opening (door) EF. There would be no difference between doors and windows (Fig. c). To get to the other side of the line GH the Flatlander would have to move around one end of the line. The line would form an insuperable barrier to him, he could not cross it or see over or under it, just as in our space a stone wall of infinite height and depth would hinder our motion and obstruct our line of sight. However, a three-dimensional being (man) with two-dimensional powers could lift the Flatlander up out of the plane, carry him across the line and then set him down again in the plane. To his Flatlander neighbors the Flatlander during this operation would suddenly vanish and remain invisible until he reappeared in their world on the other side of the line. This performance would appear as a miracle to the Flatlanders because not only would it be impossible for them to see anything not in their plane but all their experiences being confined to that plane they could not possibly explain the phenomenon. In the same manner the Flatlander might be placed in or taken out of a closed room (Fig. d) to the utter mystification of his friends. A robber from our space with two-dimensional powers could therefore enter their houses and steal their money and jewelry from their safes without breaking locks or doors. There would be nothing in Flatland corresponding to our tubes and it would be impossible to tie knots or thread needles there. As a Flatlander could not "flip" a coin, to him it would always be "heads-up" or always "tails-up." Story books have been written about life in Flatland, supposedly by Flàtland authors, in which the whole gamut of our human experiences are duplicated in so far as the restrictions under which they are supposed to live will allow. There are descriptions of life and industry in Flatland as well as tales of love and war, and the love stories close with the regulation "they married and lived happily ever after." "An Episode in Flatland" and "Scientific Romances" by C. H. Hinton; "The Fairyland of Geometry" by Simon Newcomb; "Flatland" by E. A. Abbott, which has also been dramatized and acted.

LINELAND

Let us now consider Lineland (one-dimensional space) and its supposed inhabitants, the Linelandèrs.

A Linelander (as C) would thruout life be restricted to his line (of which AB is a part) which he would conceive of as extending indefinitely in both directions. He could slide freely back and forth along this line but it would be impossible for him to move out of it. He would not be conscious of nor could he conceive of a second dimension; the concept of a plane surface would lie entirely outside of his mental realm, just as the existence of our space would be inconceivable to a Flatlander. Bodies in Lineland would not be bounded by surfaces as in our space or by contours (perimeters) as in Flatland, but by two points, their two extremities. If a Linelander encountered another similar being (as D) in his space neither could pass the other for to do so would mean that at least one of them would have to move out of the line, i.e., out of their world. Thruout life a Linelander (as C) could have only two neighbors (as D and E), one on each side of him. His experiences would be extremely limited, to human beings life in Lineland would be, to state it mildly, very dull indeed.

Let us now suppose that one of these Linelanders (as C) in some way becomes conscious of the existence of a second dimension and somehow becomes endowed with the power to move out of the line AB into a plane containing that line as, for instance, the plane of page 41. The instant he moved out of the line AB (his world) he would become invisible to his Lineland neighbors D and E The vanishing of C, apparently into nowhere, would be utterly incomprehensible to D and E; they could not account for it in any way, it would appear to them as a supernatural phenomenon, i.e., as a miracle. Possessing the freedom to move about in the plane (his new world) the Linelander C could then move back into the line AB (his old world). If he should return to a point in the line between his old Lineland neighbors D and E his appearance would be in the nature of an apparition to both of them. His reappearance, apparently out of nowhere, would be as deep a mystery to them as was his disappearance.

The happenings described in the last paragraph may also be imagined as occurring when the Linelander C has himself no conception of a second dimension and has no power to move out of the line AB (his world). We may conceive of a situation in which the Linelander C thru some influence is in the power of a Flatlander or other higher-dimensional being. This latter higher-dimensional being might then take the Linelander C out of the line AB and replace him at will, and the attendant phenomena would appear the same as before to the two Linelanders D and E. While this was taking place the Flatlander, or other higher-dimensional being who was exercising his power over C, would be invisible to the Linelanders D and E, or, if not invisible, he would appear to them only as a one-dimensional being like themselves.

A Linelander, therefore, who in some manner

has strayed off his line (one-dimensional space) into a plane (two-dimensional space) will remain invisible to his Lineland neighbors until he returns to his own world (the line).

Similarly, a Flatlander who has moved out of his plane (two-dimensional space) into space of three dimensions (our space) will remain invisible to his Flatland neighbors until he returns to his own world (the plane).

Continuing, the next step, as suggested by analogy, would then be the following. A man (three-dimensional being) who has been translated from our space into a higher-dimensional space will remain invisible to earthy beings until he returns again to our space. We have Bible records describing the bodily disappearance of human beings from our space in a manner analogous to the above. ' One case is that of Enoch who did not see death but was taken bodily out of our world by God. Gen. 5:24, Heb. 11:5. Another example is the bodily translation of Elijah into heaven. II Kings 2:11. This again suggests that space of four dimensions may be our heaven. Or, since the plural form of the word heaven occurs very often in the Bible, it may be that "the heavens" include not only space of four dimensions but also all the other higher-dimensional spaces. We have no record of a reappearance in our space of Enoch, but Elijah (and Moses) did appear to the disciples at the transfiguration of Christ. Math. 17:3, Mark 9:4, Luke 9:30. That neither Enoch nor Elijah did possess the inherent power to translate themselves from earth to a higher realm is evident from the Bible statements that they were taken by God (the highest-dimensional being). On two different occasions before his crucifixion Christ passed unseen thru threatening multitudes, since his time had not yet come. John 8:59, 10:39. Did he make himself invisible each time by entering a higher space? Was such a space Christ's abode in the interval between his resurrection and ascension and during which time he was seen in bodily form by his followers on ten recorded occasions? Each time both his appearance and his disappearance were miraculous, each one partook of the nature of a visit from a higher to a lower realm, from a higher to a lower order of space.

The Bible abounds with numerous accounts of celestial visitors (angels, archangels, prophets, and God himself) revealing themselves to inhabitants of our earth. In view of our discussion it is near at hand for us to think of them as coming down from the higher dimensional spaces in which they dwell and, after having fulfilled their missions in our three-dimensional space, to again ascend on high. Did Paul have a higher-dimensional space in mind when he wrote, "I knew a man, whether in the body, or out of the body, I cannot tèll, how that he was caught up into paradise, and heard unspeakable words." II Cor. 12:2, 3. Read again the Revelations of St. John, read them with the added light of the mathematical hypothesis of higher spaces. It will be found that much of the fog with which our limited space perceptions has enveloped it has vanished and many passages will take on a new meaning; the eyes of the soul will penetrate deeper than ever before into the dim and distant vistas which veil for us the present as well as the great unknown hereafter.

CHAPTER V

MOTION OF A POINT PERPENDICULAR TO EACH SPACE

In Chapter II the concept of the motion of a point perpendicular to a unit in each space was partially developed. We will now lead up to an extension of this idea by making use of the two following obvious propositions.

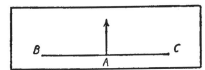 

(a) *A point (as A) starting from the center of a line (as BC) and moving off in a direction perpendicular to the line will never approach any portion of the line and will move away at the same rate from its two end points B and C.*

(b) *A point (as A) starting from the center of a circle and moving off in a direction perpen-*

*dicular to its plane (as BC) will never approach
any portion of the circumference of the circle
but will move away at the same rate from all
points on this circumference.*

Advancing another step we get:

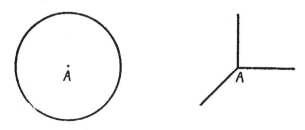

(c) *A point (as A) starting from the center
of a sphere and moving off in a direction perpendicular to our three-dimensional space will never
approach any portion of the surface of the sphere
but will move away at the same rate from all
points on this surface.*

This proposition in the Geometry of Four
Dimensions was first proven by the great astrono-
mer and mathematician Simon Newcomb, and it
reveals some of the things, extraordinary to us,
which may be done in four-dimensional space if
such a space exists. We may also state the proposition as follows. Just as in Flatland a point can
move in a direction perpendicular to a line, a
motion inconceivable to Linelanders, and in our
space a point can move in a direction perpendicular to two mutually perpendicular lines, a motion

inconceivable to Flatlanders, so in space of four dimensions a point can move in a direction perpendicular to three mutually perpendicular lines, a motion which is inconceivable to us. Assuming that the above sphere is replaced by a material hollow sphere (as a hollow sealed glass globe) and the point A is replaced by a grain of wheat, the above proposition asserts that in space of four dimensions the grain of wheat may be placed inside of the globe, or transferred from the inside to the outside of the globe, without breaking the globe or interfering with its structure in any manner. This would mean that in space of four dimensions any of our material objects can be taken out of, or placed inside of, one of our closed rooms or boxes without penetrating the walls (sides). A four-dimensional robber, therefore, could enter our houses without hindrance and steal the money and jewelry in our safes without breaking locks or doors, and none of our jails could hold him if caught. This may be where the money we miss has gone, and perhaps the articles we so often lose have accidentally rolled into four-dimensional space! A four-dimensional man could eat one of our soft boiled eggs for breakfast without breaking the shell and he could drink from one of our bottles without removing the cork. He could extract the pulp and nut from one of our plums without breaking the skin of the plum; in fact, he could remove the kernel from the nut without breaking the skin of the plum or the shell of the nut. He could take off his stockings without removing his shoes, and he could take out the inner tube of one of our automobile tires, repair it and then replace it, without removing the outer casing. Just as one of our surgeons could operate on a Flatlander without making an incision in his perimeter so a four-dimensional surgeon could operate on one of us without cutting us open. It is quite certain that if such a surgeon should locate among us he would very soon have a large and lucrative practice! All of our space, including the interior of the densest solid, is open to inspection and manipulation from the fourth dimension.

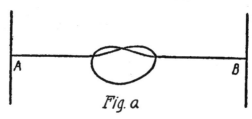

Fig. a

An ordinary knot (Fig. a) cannot be untied in our space if the two ends of the cord are fixed (as at A and B). In four-dimensional space, however, this can be done without unfastening the ends, in fact, a knot could be tied or untied on an endless string. Our knots would therefore be useless in four-dimensional space. Evidently our chains would also be useless because the links in them could be separated without breaking. Corresponding to this in Flatland would be a cord fastened at A and B (Fig. a) and making a loop around, say, C. This loop could not be unwound by a Flatlander, but one of us could do it by simply lifting that part of the cord composing the

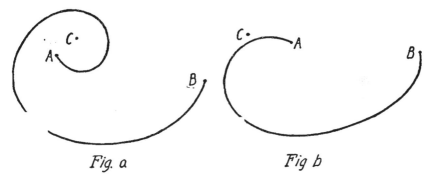

Fig. a   *Fig b*

loop into the third dimension, stretching it out, and then replacing it in the plane (Fig. b).

As has already been stated, it has been mathematically proven, beyond the shadow of a doubt, that if there exists a space of four dimensions then in that space material objects of our space may be placed in or taken out of any of our hermetically sealed closed compartments without hindrance. It has never been satisfactorily proven, however, that this can be done by a human being; our nearest approach to it is to *think* of it being done. We can *think* of the kernel of a nut being extracted without breaking the shell of the nut, in spite of the fact that we do not understand *how* it can be done. We can *think* of ourselves as entering or leaving closed compartments; their material walls do not in the least hinder our *imagining* ourselves doing it. In the Bible, however, we have records telling of two distinct occasions when Christ in the body did appear before his disciples assembled behind closed doors for fear of the Jews. John 20: 19-23, 26-29. On the first appearance Thomas was absent but on the second he was present and we read "then saith he to Thomas, reach hither thy finger, and behold my hands; and reach hither thy hand and thrust it into my side; and be not faithless but believing. And Thomas said unto Him, my Lord and my God." That Christ appeared to his disciples on this remarkable occasion clothed in his material body is further emphasized in Luke 24: 36-43. "Behold my hands and my feet that it is I myself: handle me, and see; for a spirit hath not flesh and bones, as ye see me have. . . . And they gave Him a piece of a broiled fish, and of an honeycomb. And he took it, and did eat before them." Christ, considered as a higher-dimensional being, certainly had the power to appear in his body as described above, or to do anything else which cannot be done by us in our space of three dimensions but which is possible in our hypothetical space of four dimensions by those who may dwell there. The following remarks on John 20:19 are quoted from a Pulpit Commentary on St. John. "It is more than possible—nay, it is entirely presumable—that the spiritual body becomes possessed of additional senses, of which we have no conception or experience; and, therefore, the spirit clothed with such body is alive to properties of matter and dimensions of space and active forces all of which would be supernatural to us. Our Lord before his passion gave numerous proofs of the dominance of his spirit over the body; his repeated escapes from his enemies, his transfiguration glory, his superiority to gravitation, in walking upon the sea and hushing its storms. So that he, on this occasion, is revealing to the world some of the functions of spiritual corporeity. He is manifesting the kind of life which will eventually be the condition of all the redeemed."

CHAPTER VI

ROTATION OF SYMMETRICAL CONFIGURATIONS

Let ABC and A'B'C' be two lines in Lineland LL', the points A,B,C, and A',B',C' respectively

being arranged in the same order and so located that AB = A'B', AC = A'C', BC = B'C'. One line may then be made to coincide with the other so that A coincides with A', B with B' and C with C' by simply sliding one of them along LL' This could be done by a Linelander.

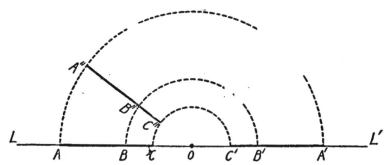

Let us now consider the two lines ABC and C'B'A' where, as before, AB = A'B', AC —

A'C', BC = B'C', but in which the corresponding points are arranged in the reverse order. In
this case the one line cannot, by sliding it along
LL', be made to coincide with the other so that A
shall coincide with A', B with B', and C with C',
and a Linelander could not conceive of any other
way in which it might be done. A Flatlander,
however, or one of us, would at once observe that
the corresponding points could be brought into
coincidence by rotating one of the lines about O in
the plane of the page, O being the center of CC'
The points A, B, C and A', B', C' respectively are
said to be symmetrical with respect to the point
O. Roughly stated, the line ABC may be taken
up into Flatland, turned over, and put down again
on the line C'B'A' so that the point A coincides
with A', B with B' and C with C'. While this
rotation was taking place the moving line, as in
the position A"B"C", would be invisible to the
Linelanders living in LL' because it would be
outside of their world; the whole process would
appear to them as a miracle.

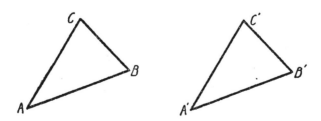

Let ABC and A'B'C' be two triangles in the plane of this page, such that AB = A'B', BC = B'C', CA = C'A', the corresponding sides of the two triangles being arranged in the same order. One triangle may then be made to coincide with the other by simply sliding it along in the plane of the page until A coincides with A', B with B', and C with C'. This could be done by a Flatlander.

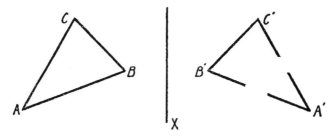

Let us now consider the two triangles ABC and A'B'C', in which as before AB = A'B', BC = B'C', and CA = C'A', but with their corresponding sides arranged in the reverse order. In this case the one triangle cannot, by sliding it along in the plane, be made to coincide with the other, and a Flatlander could not conceive of any other way in which it might be done. The reader should verify the first statement in the last sentence by cutting a triangle out of cardboard to fit, say ABC, and then, by sliding it along the page, try to make it fit on A'B'C' A three-dimensional mathematician would observe, however, that the two triangles ABC and A'B'C' are symmetrical with respect to the line XY (corresponding points on the two triangles being located on the same perpendicular to XY and at equal distances from it) and that they can be brought into coincidence by rotating one (with the plane in which it lies) about the axis of symmetry XY thru three-dimensional space until it coincides with the other. That is, one triangle must be taken up into our space, turned over, and put down on the other. To Flatlanders this operation would be an unsolvable mystery.

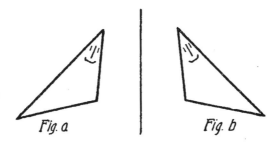

Fig. a   Fig. b

If one of us should take a Flatlander (as Fig. a) and turn him over thru our space back into his plane (as Fig. b) he would become a sort of mirror reflection of his former self. His heart would now be found on his right side instead of his left, and if he was right handed before he would now be left handed. While he was being turned over in our space he would be invisible to his Flatland neighbors; first he would apparently vanish before their eyes into nowhere and then as suddenly he would reappear like an apparition.

Before taking the next step in our discussion let us digress a trifle to point out that geometrical propositions relating to configurations in a particular space may sometimes be proven with greater ease by operating in a higher space than if we are restricted to the original space. A case in point is the plane geometry proposition that two triangles are equal if the three sides of one are equal respectively to the three sides of the other. In practically all the plane geometry text-books used in our schools the proof of this proposition requires *a rotation of one of these triangles about one of its sides.* But this means that the pupil is abandoning the space of his plane geometry to operate in the space of solid geometry (three dimensions). In like manner there are many theorems in both plane and solid geometry which can be proven with greater ease by using geometry of four dimensions than by using the methods found in our school books. In fact both plane and solid geometry have been greatly enriched by the discovery thru the geometry of four dimensions of a number of new theorems. Just as solid geometry throws a new light on plane geometry and enables us to solve problems in the plane with greater ease than if we confined ourselves to the plane, so it has been found that geometry of four dimensions illuminates both plane and solid geometry, and has made easy or rendered possible the solution of many problems both in the plane and in our three-dimensional space which, without the concepts of geometry of four dimensions, would have been very difficult of solution to say the least. Schubert has illustrated this by means of some very interesting examples.

We now return to our line of discussion.

Let ABCD and A'B'C'D' be two tetrahedrons in our space having all the faces of one equal to the corresponding faces of the other and arranged in the same order. By superimposing one on the other we can make one coincide with the other without difficulty.

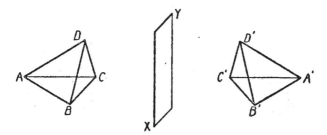

Let us now consider the two tetrahedrons ABCD and A'B'C'D' in which, as before, all the faces of one are equal to the corresponding faces of the other, but arranged in the reverse order. No amount of manipulation will now enable us to bring the two tetrahedrons into coincidence, and no human being can conceive of any way in which it can be done. This task is as impossible to us as was the task of the Linelander to bring the symmetrical lines ABC and C'B'A' into coincidence (p. 54) or the task of the Flatlander to make the two symmetrical triangles ABC and A'B'C' coincide (p. 56). Mathematicians of the highest standing, however, have proven that it can be done in four dimensional space by rotating one of the tetrahedrons about their plane of symmetry XY (corresponding points of the two tetrahedrons being located on the same perpendicular to XY and at equal distances from it), a kind of rotation that is impossible in our space of three dimensions. Or, stating it roughly, one of the tetrahedrons must be taken up into space of four dimensions, turned over, and then brought back into concidence with the other. One of these tetrahedrons corresponds to the mirror image of the other, the mirror being in the plane of XY. While this rotation is taking place the moving

tetrahedron would be invisible to us; just as the
rotating line was invisible to the Linelander (p.
54) and the rotating triangle was invisible to the
Flatlander (p. 56).

↑ When you look at your image in a mirror, the
right and left sides of your body appear to have
exchanged places. A mole on your right cheek
is seen on the left cheek of your image. Your
image is symmetrical to your body with respect
to the plane of the mirror. You may go behind
the mirror to the place where your image appeared but you cannot take the exact position of
the image, or rather, you cannot make your body
coincide in every detail with that position. Turn
and twist about as much as you please in our
space you cannot make your right and left sides
exchange places. But a four-dimensional being
could do that very thing to you by rotating you
about a plane in his space. And, what is more,
no part of your body would be dislocated during
the operation, you would not experience any tearing or wrenching of the tissues of your body.
After this "reversal" or "twist" in space of four
dimensions, during which you would be invisible
to all other human beings, everything on you
would also be changed from right to left and
vice-versa, even the seams and buttons on your
clothes, every dimple and wrinkle on your face

and every hair on your head. Your point of view would be completely turned around, so that everyone else would appear to you to be reversed. The letters on this page would seem reversed to you, i.e., as they really do appear on the printer's type form. The hands of a clock would appear to you as going backward and the sun to rise in the west and set in the east; the whole world would appear to you as a looking-glass world. When you look into a mirror you see a picture of the actual "reversal" or "twist" that can be imparted to material bodies in the space of four dimensions, but which cannot be brought about in our space of three dimensions.

Our two hands appear to be at least approximately of the same size, and shape but we can never manipulate one so that it shall occupy the same space that has been occupied by the other; we cannot put a right glove on the left hand, or vice-versa. The reason for this is that our two hands are symmetrical with respect to some plane in our space and, like the two symmetrical tetrahedrons shown on p. 59, they cannot be made to coincide by any operation entirely confined to our space of three dimensions. For the same reason we cannot wear a left shoe on the right foot, or vice-versa. A four-dimensional man, however, can change a right shoe into a left shoe by rotating it about a plane in his space. If a one armed man in our space should in some way become endowed with the power to rotate bodies about planes in four-dimensional space he could use both of an ordinary pair of gloves because he could change either one into the other, so that he would have either two rights or two lefts.

Instead of rotating material bodies about a plane in space of four dimensions as described above it has been proven that a curved surface may also be used as the "axis" of rotation by allowing for a slight degree of distortion. This leads to very interesting results if the rotation is applied to flexible material surfaces like a curved sheet of rubber. The two surfaces of the sheet of rubber may be made to exchange places by rotating them about the sheet itself in space of four dimensions. During the rotation there would be no interference between portions of the surfaces so that it may be done with surfaces that are closed (like a sealed hollow rubber ball) as well as with surfaces that are open (like a sheet of rubber).

This means that a hollow rubber ball (Fig. a) could be turned inside out without tearing just as we may turn a rubber band or the section of an inner tire tube (Fig. b) inside out. In both cases the inside and outside surfaces would exchange places.

Fig. a

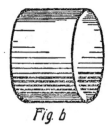

Fig. b

Not only in the animal kingdom do we find this wonderful symmetry (right and left handedness) exemplified in the bodies of all creatures, as well as in most of the interior organs of those bodies, but it also appears very often in the vegetable kingdom as shown by the symmetry of the two sides of a leaf, and in flowers, bulbs, nuts, fruits, etc. In fact, the right and left rotation (twist) which is characteristically four-dimensional is of very frequent occurrence in nature. A beam of polarized light (whose wave-vibrations are all in one plane) is rotated either to the right or to the left on passing thru certain organic substances (sugars, starches) just as we might hold a ribbon by the ends and give it a twist to the right or left. Two forms of sugar found in honey called dextrose and levulose owe their names to the fact that one rotates a polarized beam to the right (dextra, "right hand") and the other to the left (læva, "left hand"). In chemical constitution

they are exactly the same, they are both represented by the same formula, and yet they rotate
the plane of polarization of a beam of light in
opposite directions. Such substances are called
"isomeric." A molecule of each of the two substances contains the same number of each kind
of atom, the difference in the effect on light being
accounted for by a different arrangement or grouping of the atoms in the two molecules.

The following discussion will explain what relation this has to a possible four-dimensional space.

In a plane three points (as the vertices of the

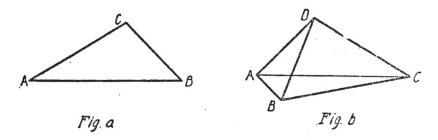

Fig. a   Fig. b

triangle in Fig. a), and no more, can be located
so that the three distances between them shall
be independent of each other, i.e., any one of the
distances may change without changing the others.

In our space of three dimensions four points
(as the vertices of the tetrahedron in Fig. b), and
no more, can be located so that the six distances
between them shall be independent.

In space of four dimensions five points, and

no more, can be located so that the ten distances between them shall be independent, something that cannot be done in our space.

If then a molecule consists of five atoms we cannot in our space alter the distance between two of them without at least altering some second distance. "But if we imagine the centers of the atoms placed in four-dimensional space, this can be done; all the ten distances which may be conceived to exist between the five points will then be independent of one another. To reach the same result in the case of six atoms we must assume a five-dimensional space; and so on. Now, if the independence of all possible distances between the atoms of a molecule is absolutely required by theoretical chemical research, the science is really compelled, if it deals with molecules of more than four atoms, to make use of the idea of a space of more than three dimensions. This idea is, in this case, simply an instrument of research, just as are, also, the ideas of molecules and atoms—means designed to embrace in a perspicuous and systematic form the phenomena of chemistry and to discover the conditions under which new phenomena can be evoked. Whether or not a four-dimensioned space really exists is a question whose insolubility cannot prevent research from making use of the idea, exactly as chemistry has not been prevented from making use of the notion of atom, altho no one really knows whether the things we call atoms exist or not." (Schubert.)

For example, at least eight different alcohols have been discovered having the same formula $C_5 H_{12} O$. Chemists account for these different compounds by supposing that there is a different arrangement or grouping of the five carbon, or C-atoms, in each compound, something which would be possible in space of four dimensions but not possible in our space. In other words, should it become necessary for the explanation of the structure of this molecule to assume that the five carbon atoms in it are equidistant from each other (the simplest possible case) we would also have to assume the existence of four-dimensional space, because in that space, and not in our space, is it possible to locate five points equidistant from each other.

There are two varieties of tartaric acid which crystallize into symmetrical forms, forms bearing the relation of object to mirror-image as here shown.

Investigators have found that one of these crystals apparently changes into the other without chemical resolution and reconstitution and without any manifestation of force. If it could be proven without the shadow of a doubt that this is what actually happens, then this would point to the existence of space of four dimensions, because in that space, and not in ours, is it possible for a right-handed shape to change into its symmetrical left-handed shape by a simple movement. Certain snails, exactly alike in other respects, differ similarly in that some are coiled to the right and others to the left. An astonishing phenomenon connected with these snails is the fact that their juices have the property of rotating a polarized beam of light to the right or to the left corresponding to the direction in which the snails themselves are coiled. This suggests that their external form is the expression of an internal difference due to a right or left twist of their atoms caused by a four-dimensional force. It has been suggested that such a four-dimensional twist runs thru all living forms, and that the life force is, in part at least, four-dimensional.

CHAPTER VII

SECTIONS OF BODIES IN ONE SPACE MADE BY SPACES OF A LOWER ORDER. CONSERVATION OF MATTER AND ENERGY

Suppose we pass a circular cone downward perpendicularly thru Flatland XY (see figure). The cross sections of the cone made by the plane of

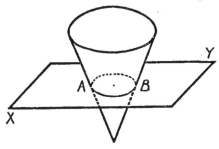

Flatland will be circles (as AB). The Flatlanders living in the plane would see at first only a point as the apex of the cone entered their plane, then they would see a circle gradually increasing to the size of the base of the cone, when suddenly every trace of the cone would vanish. The Flatlanders would have no conception whatever of the cone as we know it, indeed, they would not suspect that the two-dimensional object (circle)

which they had seen had anything to do with a
three-dimensional solid at all, even if they had an
inkling of the existence of our space. The phenomenon would appear to them as the creation of
a new two-dimensional body (circle) which continued to grow larger and larger and then suddenly vanished. The alpha and omega of this
phenomenon would be deep mysteries to them,
just as birth and death are to us. In like manner
it may be that each one of us is in reality a four-
(or higher-) dimensional being, our personality
(as we are known to ourselves and others) being
the section of our real, our higher self, made by
the three-dimensional space of our experience.
Like the cone passing thru Flatland, we arrive into
this three-dimensional world from some unknown
shore, i.e., are born; we grow up thru childhood,
youth, middle age, old age; we die, and our souls,
our real, our higher selves, pass into the great
unknown hereafter. And thruout life we have
been thought of by ourselves and our fellowmen
as three-dimensional beings only. Perhaps all living things, as they appear to us, are the sections
of four- (or higher-) dimensional creatures made
by our space. "Viewing human life from this
standpoint, the conclusion may be reached that I,
as I write this, am merely that section of my four-
dimensional self that happens to be passing thru

this world at this moment, and that the whole of me, from my birth to my death, is a four-dimensional entity; that the past and the future are past and future only in a three-dimensional sense, and that in a four-dimensional sense the past and future are present—that is, both what was and will be, is." (Taylor.)

Suppose now that a particular Flatlander had in some way acquired the intellectual power to sum up, in his mind, in a moment of time, all the cross sections (circles) of the cone he had seen. He would then obtain a mental picture of the cone, not as we see it, but as a body, not of two dimensions, but as a body having three dimensions, time being the third dimension. In like manner, by a concentrated mental effort, we may in a moment of time assemble all the impressions we have received of some friend or relative with whom we have been intimately associated from his, or her, birth to death. The mental image thus obtained would include all the years of his or her life, the whole span of life of that individual would be compressed into a moment of time; and this mental image would have four dimensions, time being the fourth dimension. Mental vision has therefore been called four-dimensional by some writers. All of us have in our dreams experienced this crowding together of a long series of occurrences into an instant of time. There are no restrictions of time, or restrictions of the space of our experience, in dreamland. "The unthinkable` velocity of time in dreams may be inferred from the fact that between the movement of impact of an impression at the sense periphery and its reception at the center of consciousness—moments so closely compacted that we think of them as simultaneous—a coherent series of representations take place, involving what seem to be protracted periods for their unfoldment. Every reader will easily call to mind dream experiences of this character, in which the long-delayed dénouement was suggested and prepared for by some extraneous sense impression, showing that the entire dream drama unfolded within the time it took that impression to travel from the skin to the brain." (Bragdon). Perhaps dreamland is located in four-dimensional space.

A man who has experienced the sensations of a dying person as, for instance, a man who believed he was drowning, tells us that in an instant the whole panorama of his past life was spread out before him in a vivid momentary perspective. "Memory is a carrying forward of the past into the present, and the fact that we can recall a past event without mentally rehearsing all the intermediate happenings in inverse order, shows that in the time aspect of memory there is simultaneity as well as sequence." If our mental vision is four-dimensional it points to the possibility that our mental or spiritual self is four- (or higher-) dimensional. If time is but the way in which we perceive the fourth dimension, then our spiritual selves, being higher-dimensional, are independent of time, above time, outside of time, that is, eternal. In this higher space our grammatical tenses would then have no meaning. Perhaps this explains why these tenses as used in the Bible sometimes appear "mixed" to us. Jesus said "before Abraham was, I am." "For a thousand years in thy sight are but as yesterday when it is past." Ps. 90:4. That God, the highest-dimensional being is not, like we, restricted by time limitations is clear. "One day is with the Lord as a thousand years, and a thousand years as one day." II Peter 3:8. "Behold, thou hast made my days as an handbreadth; and mine age is as nothing before thee." Ps. 39:5.

Another interesting illustration similar to that of the cone passing thru Flatland (p. 69) is that of a wire bent in the form of a circular helix (coiled wire spring) which is passed thru Flatland so that its axis is perpendicular to the plane of Flatland (see figure). The cross section of the wire made by the plane (as A) will be a very

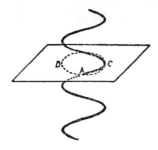

small ellipse (two-dimensional). If the wire is passed downward the Flatlanders would see this ellipse travel in the circle BC in one direction, while if the wire is passed upward it would travel in the opposite direction. The "birth" of this ellipse as the wire entered the plane, its "death" when it left the plane, and its behavior in the interim—all would be inscrutable mysteries to the Flatlanders living in that plane.

It is evident that if the cone to which we referred on p. 69 is passed obliquely thru the plane, then the sections will no longer be circles but ellipses, parabolas, hyperbolas, or straight lines, all depending on the angle which the axis of the cone makes with the plane. To further illustrate that the *kind* of section made in a given case will depend not only on the *shape* of the solid but also on *how* the solid enters the plane, let us assume that the same cube passes thru the plane several times, each time with a new orientation with respect to the plane.

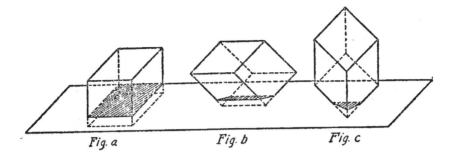

Fig. a   Fig. b   Fig. c

If a face of the cube is always parallel to the plane. then all the sections made will be equal squares (Fig. a).

If an edge of the cube (and no face) is always parallel to the plane then all the sections made will be rectangles except the first and last which would be lines, i.e., the top and bottom edges (Fig. b).

If no edge or face is parallel to the plane, i.e., the cube is passed thru obliquely, the first section will be a point (a corner of the cube), then they would be triangles, then polygons, then again triangles, and finally the last one would again be a point (Fig. c).

Since we live in space of three dimensions the above phenomena are fully understood by us, but to Flatlanders viewing the appearance, behavior, and disappearance of these sections in their plane, it would all be utterly inexplicable because they have knowledge and experience only of two-dimensional bodies. Least of all would they suspect that sections so unlike were produced by the passing of the same higher-dimensional body thru their world.   If then, as it has been suggested, the personality of a man is the section of his real, his higher-dimensional self, made by the space of our experience, it makes it clearer to us why the personalities of all the people we know have such characteristic differences.   Both as to mind and body they all seem to have been cast in different moulds.   All this suggests that the personality of a man may depend not only on the *character* of his higher-dimensional self but that it is also influenced by the *manner* in which that higher self was projected into our world.

In the preceding paragraphs of this chapter the reader naturally thought of the cone, helix, and cube as substantial solids, while the sections (points, circles, triangles, lines, etc.) of these solids made by the planes thru which they were passing impressed him with their changing, fleeting, and evanescent character.   Passing now to the next higher order of space, analogy suggests the above relations as one explanation of the transitory character of our lives; they may be merely the sections of our spiritual, our higher-dimensional selves made by the three-dimensional space of our experience.   Many passages in the Bible emphasize the ephemeral character of our earthly existence, while others point out the higher, the divinely transcendental origin and nature of man. "As for man, his days are as grass; as a flower of the field, so he flourisheth. For the wind passeth over it, and it is gone; and the place thereof shall know it no more." Ps. 103:15, 16. "For what is your life? It is even a vapour, that appeareth for a little time and then vanisheth away." James 4:14. "Man that is born of a woman is of few days, and full of trouble. He cometh forth like a flower, and is cut down: he fleeth also as a shadow, and continueth not." Job 14:1, 2. "Mine age is departed, and is removed from me like a shepherd's tent." Is. 38:12. "So God created man in his own image, in the image of God created he Him." Gen. 1:27. "For thou hast made him a little lower than the angels, and hast crowned him with honor and glory." Ps. 8:5. "For in him we live and move and have our being." Acts 17:28.

In this connection it is of interest to refer to the method employed by the biologist to determine the interior structure and composition of solids. For instance, by cutting an apple into a large number of thin slices of uniform thickness and arranging them in the proper sequence on a flat surface a clear concept of the structure of the apple may be obtained. That is, he mentally sums up the total of the impressions on his mind made by these slices and thus obtains a mental image of the interior as well as the exterior of the apple. Thus, if the apple happens to be worm-eaten it is easy to trace the passage made by the worm, as well as to locate the worm, by noting the relative locations of the series of worm holes in the slices. This method is of the greatest utility to the biologist in the study of the structure and composition of the embryos or organs of animate beings. For instance, the embryo of a frog is cut into a large number of sections by means of the microtome, an instrument constructed for this purpose, so thin that they are practically transparent under the microscope. These slices, which are practically plane sections of the embryo, are then mounted on glass plates by means of a perfectly transparent adhesive and studied under the microscope. These plates, called slides, when consecutively arranged in a series, will vividly protray the most minute detail of the structure of the embryo. By means of these slides the biologist is able to mentally reassemble all these sections in their proper order and thus to mentally reconstruct the embryo.

CONSERVATION OF MATTER AND ENERGY

Again making use of a cube passing obliquely downward thru a plane (as in next figure) we note

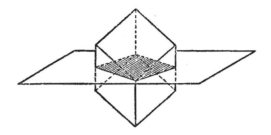

that the section of the cube made by the plane is continually changing in magnitude. Starting as a point (when the lowest corner of the cube enters the plane) the section changes into an increasing triangle, then into polygons of more than three sides, then into a decreasing triangle which finally contracts into a point (as the highest corner of the cube emerges from the plane).

The Flatlanders living in that plane would observe that the area of this section is continually changing, i.e., this area does not remain constant. This suggests the possibility that the total mass of our material universe (considered as the section of a higher-dimensional configuration passing thru our space of three dimensions) may in like manner be continually changing, i.e., the total mass of our material universe may not be constant and hence the principle of the conservation of matter may not hold true. And this may apply not only to our material universe as a whole but also to all of its parts from the smallest atom to the largest planets and suns. To illustrate again, suppose that instead of the cube an automobile is passing downward thru the plane. The plane section of the automobile will be of very great complexity, and not only the whole of the section but also all of its many parts would be continually changing in form and area. And not only that, but during the interval between the instant when a section of the automobile first appeared in the plane and the instant the last trace of it vanished the particular sections of hundreds of the individual parts of the automobile would suddenly appear (be created), pass thru a continuous series of changes, and then as suddenly vanish (be annihilated). For instance, we can visualize the section of some particular bolt in the automobile as it is passing thru the plane.

May it not be possible that the whole of our material universe as well as all its component parts are similarly changing in form and total mass so that there may be additions to or subtractions from this total which we as three-dimensional beings cannot account for? Astronomers have observed the sudden appearance in the heavens of stars and comets, and also the apparent vanishing of such celestial bodies, all in a manner suggesting that they were either entering or leaving our universe. For these and other reasons there has been for some time a suspicion in the minds of many scientists that the total mass of our material universe is not constant. Since every atom of matter is a reservoir of energy it follows that if the principle of the conservation of matter is not true, then the principle of the conservation of energy also fails. Numerous instances are recorded in the Bible where new matter or new energy was apparently added to our material universe by supernatural means. The manna falling in the wilderness, the ever replenished oil in the widow's cruse, and the loaves and fishes which fed the five thousand, all this food may have been the three-dimensional sections of higher-dimensional supplies projected at those particular times thru our space by some higher agency. Also the energy which divided the waters of the Red Sea, which caused the walls of Jericho to fall, and by means of which Jesus stilled the storm at sea and walked on its waters, this energy was in some to us mysterious way added to the already existing energy in our universe. When God translated Enoch and Elijah bodily into heaven matter and energy was subtracted from our material universe.

CHAPTER VIII

THE ILLUSORY AND THE REAL

In the case of the cube passing thru a plane it is evident that the cube itself is the real, the true individual, while the sections (squares, triangles, etc.) of it seen by the Flatlanders living in that plane are but illusory, inadequate images of the cube. However, it is only thru these countless sections that the cube is manifested to the Flatlanders. But the "birth" of a section when the cube enters the plane is not the "birth" of the cube, and the "death" of a section when the cube leaves the plane is not the "death" of the cube; this "birth" and this "death" are mere illusions as far as the cube itself is concerned. And so, because man is a higher-dimensional being his birth, that is, his entrance into this world, is not his beginning, nor will his death, that is, his passing out of this world, be his end.

The classical allegory of the chained shadow-watchers given by Plato in the seventh book of his "Republic" vividly portrays the relation between true being and the illusions of the sense world. Plato imagines a group of prisoners chained just inside the entrance to a large cavern. All movement is impossible to them, they cannot even turn their heads, hence they always look upon the opposite wall of the cavern. A strong light streams in thru the mouth of the cavern. The prisoners never see anything but their own shadows on the wall together with the shadows of all persons and objects moving behind them. In the course of time they would identify themselves with their shadows, they would refer all their experiences to these shadows, they would practically become the denizens of a shadow world. All movements observed by them would be the movements of shadows on the wall, the movements of shapes with outlines but with no substantiality. Plato thus conceived of a possible state in which man is reduced by his limited consciousness to less than he really is and thus cleared the way for the concept that the normal man in an analogous manner is reduced by *his* consciousness to less than he really is. Man is thus but the shadow cast on our world by his real, his higher self dwelling in supernal regions. As the real world of the chained prisoners was to their shadow world, so is the supernal world to the world of our experience.

Let us suppose the surface of a smooth wet

sidewalk to be part of Flatland.   A man walking on it would make tracks as shown above.   The Flatlanders living in the plane of the surface of the sidewalk would see only the bottoms of the heels and soles of his shoes.   First the heel and sole of one foot would appear to them apparently out of nowhere and then vanish just as the heel and sole of the other foot appeared and in turn vanished.   This would be repeated again and again to the great mystification of the Flatlanders. To them it would seem as if two two-dimensional bodies (the bottoms of the heel and sole of one shoe) were traveling together for some unknown reason accompanied by their twins (the bottoms of the heel and sole of the other shoe).   The Flatlanders could not conceive of each heel and sole being connected, that is, as belonging to the same shoe (a three-dimensional body), and much less could they conceive of these shoes as belonging to the same three-dimensional being (the man).   Suppose, however, that some Flatlander of superior intellect had an inkling of the third dimension such as we have of a fourth dimension, and that he suspected that the two pairs of impressions made on the sidewalk really belonged to the same three-dimensional being. As only the bottoms of the man's heels and soles would come within his range of observation he would probably state as his opinion that this three-dimensional being (the man) had no intelligence or feeling and therefore was of little or no account. He would judge the man from what he had seen of the bottoms of his heels and soles! By analogy then, if man is a higher-dimensional being of whom we can observe his three-dimensional attributes only, it may be that, comparatively speaking, we have no clearer or fuller conception of what man actually is in the sight of God than had the Flatlander when he tried to describe a three-dimensional being (the man) from what he had seen of the bottoms of his heels and soles.

CHAPTER IX

SPACES AS HEAVENS AND HELLS

Perhaps the higher-dimensional spaces are the heavens to which we ascend at death if we have so lived on earth as to deserve such promotion; and the lower-dimensional spaces are the hells to which we will be consigned in case our lives have fallen below a certain standard. Thus at death the soul of a good man (his higher-dimensional self) shakes off the limitations of understanding, knowledge and movement which our three-dimensional space imposed on him, and enters space of four dimensions. Instead of being restricted to three degrees of freedom of motion he will then be able to move in four independent directions. Such would be his increased freedom of action that, if he could look back, his former life in our space would probably appear to him as life in a prison cell would now seem to us. Unlimited vistas of new knowledge would open up to him, and his powers of understanding and possibilities of action would have increased to an extent far beyond the present limits of human conception.

86

Coming from the restrictions imposed by our space, the space of four dimensions would indeed seem heaven to him! On the other hand, the soul of the wicked man would at death be condemned to live in Flatland, the space of two dimensions. He could move only in two independent directions, all his knowledge and experience of a third dimension would be blotted out of his mind, and life in Flatland would certainly be hell to him. If during his existence in Flatland the fallen man did not mend his ways but continued to deteriorate in character, then at death he would again be demoted, he would be consigned to Lineland, an experience that would abound with horrors. He could then move in one direction only and his fund of knowledge and his powers of understanding would be still further diminished. If after this awful experience he still persisted in his evil ways he would at death be consigned to space of zero dimensions, that is, to a point. There no motion at all would be possible, all his knowledge and understanding would have vanished, he would be in the deepest pit of hell. "For a fire is kindled in mine anger, and shall burn unto the lowest hell." Deut. 32:22.

But let us turn back to the more pleasant prospect, the assumption that our earthly friend has so lived that at death he is translated into space of four dimensions, that is, the first heaven for us. "And I saw a new heaven and a new earth: for the *first heaven* and the first earth' were passed away." Rev. 21:1. Happily during his life in space of four dimensions our friend continues his striving after that which is highest and noblest and best, and so when he has finished his course there he is promoted to space of five dimensions, that is, our second heaven, a place of far surpassing knowledge and freedom and glory. Continuing in growth of character he would then be advanced to space of six dimensions, that is, our third heaven. "I knew a man in Christ above fourteen years ago (whether in the body I cannot tell; or whether out of the body, I cannot tell: God knoweth), such an one caught up into the *third heaven*." II Cor. 12:2. Not stopping here we can imagine that the noblest and greatest individuals will, after such successive existences, be translated to still higher spaces (heavens). The higher the order of space the greater and more glorious would be the state of the blessed who dwell there, until the space of infinite dimensions is reached where God has his dwelling place. *There* are no restrictions to motion and no limitations of power, knowledge or understanding. God is omnipotent, omniscient, omnipresent; the very personification of justice and mercy, wisdom and love. "God is absolutely infinite, consisting of infinite attributes each expressing eternal and infinite essentiality."

The Bible has made the above concept of more than one heaven familiar to us all. "In my Father's house are many mansions." John 14:2. "The heaven, even the heavens, are the Lord's." Ps. 115:16. "When I consider thy heavens." Ps. 8:3. "When he prepared the heavens." Prov. 8:27. "Therefore will I shake the heavens." Isa. 13.3. "For, behold, I create new heavens." Isa. 65:17. Also Hag. 2:6, Ezek. 1:1, Matt. 3:16, Mark 1:10, II Cor. 5:1, II Peter 3:13. We have referred to higher and lower orders of spaces. That there are also various orders of heavens may be inferred from scripture passages. "Behold, the heaven and the heaven of heavens is the Lord's thy God." Deut. 10:14. "Behold, the heaven and the heaven of heavens cannot contain thee." I King 8:27.

We find nothing in the hypothesis of higher spaces that we have been developing which contradicts the reality of sin or the fact of the divine redemption of a fallen race. Whether or not the inhabitants of other spaces than ours have passed or are passing thru the experiences of an Adam's fall and a Christ's atonement we do not know, but the presumption is strong that at least some of the dwellers in higher space (heavens) did disobey God and therefore incurred the inevitable punishment which is sure to overtake the sinner, whether he be man or angel. "Behold, he putteth no trust in his saints: yea, the heavens are not clean in his sight." Job 15:15. "For if God spared not the angels that sinned but cast them down to hell, and delivered them into chains of darkness, to be reserved unto judgment." II Peter 2:4. "And the angels which kept not their first estate, but left their own habitation, he hath reserved in everlasting chains under darkness unto the judgment of the great day." Jude 6. "And he said unto them, I beheld Satan as lightning fall from heaven." Luke 10:18.

CHAPTER X

CREATION OF OUR MATERIAL UNI- VERSE. EVOLUTION

Is the theory of higher spaces in accord with the Bible account of creation? Let us approach this subject by making use of the concept employed in the last chapter, the assumption that bodies in any given space are the sections of higher-dimensional bodies made by that space. For example, let the plane AB represent a portion of a two-dimensional space (Flatland).

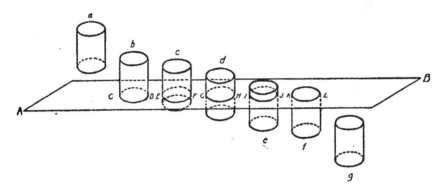

Assume figure (*a*) to be a solid circular cylinder (three-dimensional body) with its axis always perpendicular to the plane and moving towards and completely thru the plane and assume the rest of the figures as successive stages of the same cylinder as it is passing thru the plane. Before the cylinder reaches the plane, Flatlanders living in that plane would know absolutely nothing about it, the cylinder would be non-existent as far as they are concerned. The instant the lower base of the cylinder coincided with the plane, as in figure (*b*), the circle CD, the section of the cylinder made by the plane, would suddenly appear before the astonished Flatlanders, coming apparently out of nowhere. To them this phenomenon would be the *creation of a circle* (*two-dimensional body*) *out of nothing. They could no more account for this miracle than we can account for the creation of our material* (*three-dimensional*) *universe out of nothing.* They could not even imagine an explanation of the mystery, they could simply state that where there was nothing before there was now a circle; a real two-dimensional body had been created apparently out of nothing.

During the time the cylinder is passing thru their plane the Flatlanders would behold the circular sections made by the plane as a continuing entity, a circle, (as EF, GH, etc.) until its upper base KL, figure (*f*), coincided with the plane. Then the instant the cylinder passed out of the plane the circle would suddenly vanish, apparently into nowhere. To the Flatlanders the circle would cease to exist, where there had been something before there was now nothing. *They could no more account for the annihilation of the circle than we can conceive of the possibility of the annihilation of our material universe,* it would be and remain an unsolvable mystery to them.

Continuing this line of reasoning one step further we may think of our material universe as the section of a higher-dimensional body (aggregate) made by our space of three dimensions. Just as the circle, in what has gone before, was a two-dimensional body created by the entrance of a cylinder (three-dimensional body) into two-dimensional space (Flatland) so we may conceive of our material universe as being created by the entrance of a higher-dimensional body (aggregate) into our three-dimensional space. As human beings restricted to the knowledge and experiences of a three-dimensional space we can have no knowledge of how this can be brought about or understand the real nature of the higher-dimensional body (aggregate) in question. All that we can say is that where there was nothing before we now behold our material universe, that is, our material universe was created out of nothing. "In the beginning God created the heaven and the earth." Gen. 1:1. "Through faith we understand that the worlds were framed by the word of God, *so that things which are seen were not made of things which do appear.*" Heb. 11:3.

Just as Flatlanders (p. 95) continued to behold as a continuing entity the circular section of the cylinder made by their space (plane) while the cylinder was passing thru the plane, so we today behold our material universe as a continuing entity, an aggregate of planets, stars, and nebulæ, which aggregate we may conceive of as being the section of a higher-dimensional configuration of surpassing greatness and variety formed by the passing of this transcendent configuration thru our space of three dimensions. If, in our illustration, instead of a circular cylinder some irregular shaped solid had been passed thru the plane then the section seen by the Flatlander would have been, not a circle, but a continually changing irregular shaped plane figure (two-dimensional body). And so, because every part of our material universe from the smallest particle of matter on earth to the most distant star is in a state of continual change we may assume that these changes are merely three-dimensional manifestations of a transcendental configuration of infinite richness in its composition and form caused by the passing of the latter thru our space of three dimensions. According to this view the

events of the six days of creation as recorded in
the Bible, as well as all other events which have
since taken place and are at this moment taking
place, are three-dimensional sections (images,
manifestations) of some higher-dimensional aggregate (heaven) in which God dwells.

In the case of the cylinder and the plane we
pointed out on pp. 92 and 97 that as the cylinder
passed out of the plane the circle which the Flatlanders had observed suddenly vanished before
their eyes after its coincidence with the upper base
of the cylinder, and was seen by them no more.
No trace whatever of the circle remained in the
two-dimensional world of the Flatlanders, to them
it appeared as the total annihilation of a body
(the circle) in their space (plane), a phenomenon
which they could in no way account for. The
Flatlanders might well record it in their annals
as the end of the circle. And so to us (three-
dimensional beings) the end, the total annihilation, of our material three-dimensional universe may be considered as coming to pass when
the higher-dimensional (transcendent) configuration to which we have referred passes out of our
three-dimensional space. Numerous Bible passages indicate unmistakably that not only did our
universe have a beginning (was created) but that
it will also in the fullness of time have an end

(be annihilated) Ps. 75:3, Mat. 13:40, II Pet. 3:10, 13, Rev. 21:1. The Bible does not, and could not, tell us how something can become nothing any more than it can tell us how nothing can become something because it was written for human (three-dimensional) beings wholly incapable of comprehending such (higher-dimensional) mysteries.

EVOLUTION

Let us assume a solid circular cone (three-dimensional body) passing thru a plane, as AB,

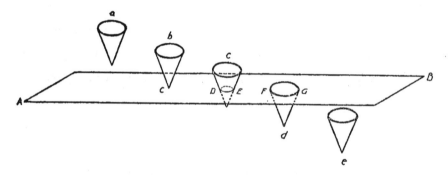

(two-dimensional space) the axis of the cone being always perpendicular to the plane.

Before the cone reaches any part of the plane, as in the position fig. (*a*), the cone is non-existent to the Flatlanders living in the plane. The instant the apex of the cone enters the plane (fig. (*b*), the point C (the section of the cone made by the plane) will appear to the Flatlanders in that plane coming apparently from nowhere. To them it will be the creation of a point out of nothing. Then as the cone continues to move thru the plane this section (point) will be seen to develop into a circle (as DE) with a continually increasing radius. In other words, the Flatlanders would behold *the evolution of a point into a circle.* And this evolutionary process would continue until the circular section has grown into the circle FG, the base of the cone. As the cone passes out of the plane the circle will vanish, apparently into nowhere. A Flatlander living in the plane AB and observing all that had taken place might well describe it all as the *creation, the development by evolution, and the annihilation of a circle;* and this without having any conception whatever of the existence of the cone itself or its relation to the phenomenon he had witnessed.

May it not be that all evolutionary processes in our material three-dimensional universe are in like manner three-dimensional sections (images, manifestations) of the elements of some higher dimensional (transcendent) aggregate of infinite complexity? In this concept there is no conflict between the doctrine of instantaneous creation and the theory of evolution, each is a part of the process involved. Both of them, however, operate in a manner incomprehensible to three-dimensional beings, both are miracles to us. To God instantaneous creation and evolutionary processes are equally possible, he can and does operate with either or both according to his infinite wisdom. To deny the possibility of instantaneous creation is to deny the power of God and to cast overboard the concept of evolution is to deny the existence of the numberless examples of evolutionary processes which are developing today before our very eyes or are unfolded to our minds.

CHAPTER XI

RAISING OF THE DEAD. SACRAMENT OF THE LORD'S SUPPER

The soul of man, being higher-dimensional, can leave the human body, which is three-dimensional, without doing any violence to the body or leaving any trace on it of a point of emergence. Investigators have often tried to photograph the departing soul of a dying man, and to discover any lesions on his body caused thereby, but always without success. No image of the soul, which is higher-dimensional, can be secured by material (three-dimensional) means; nor can the human body imprison the soul any more than (as we have pointed out) it would be possible to imprison a four-dimensional being in a three-dimensional prison cell or to confine a three-dimensional being within two-dimensional (Flatland) barriers.

Likewise it follows that a soul may enter a human body without hindrance as far as spacial considerations are concerned, just as a three-dimensional being may freely enter a closed compartment in Flatland. The raising of the dead

involves the re-entrance of souls into their former corporeal habitations. What the forces are
which can bring this about is a deep mystery to
us. We know, however, that Jesus had the power
to perform this miracle, and that thru him, also
Paul and Peter raised the dead. Jesus restored
to life the daughter of Jairus (Mat. 9:16: Mark
5:4: Luke 8:54), the son of the widow of Nain
(Luke 7:15) and Lazarus (John 11), who had
been four days in his grave. Paul did the same
for Eutychus (Acts 20:10) and Peter the same
for Tabitha (Acts 9:40). Likewise Elijah (I
King 17:17, 23) and Elisha (II King 4:32, 37)
raised the dead. The dead came out of their
graves on the first Good Friday, later entering
the Holy City and appearing unto many. (Mat.
27:52).

That we have no record of the experiences of
Lazarus and the others who were raised from the
dead, covering the interval between their deaths
and their coming to life again, has been a keen
disappointment to many Bible readers. In fact,
this seeming omission of a description of the de-
tails of their existence during this death interval
has been used by Bible scoffers not only to cast
doubt on the truth of these miracles but also to
discredit all Christian beliefs. Where was the
immortal part of Lazarus during the four days

in which his body was in the grave, what did he do, what were his experiences? A message from Mars would be of trivial interest to dwellers on earth compared with the importance of the message that Lazarus should have brought back from the regions beyond the grave. It would have answered the great burning question of past, present and future ages, the question beside which all other questions vanish into insignificance. That Lazarus did not answer this question is evident because if he had it would have been recorded. The people of his day were no less consumed by interest in the answer to this question than we are today. If Lazarus had related his experiences during those four fateful days his narrative certainly would have been handed down to us in some form.

The reason why Lazarus did not answer this question, did not relate his experiences between his death and his return from the grave, was because he could not. Granting that he had a conscious existence during this four-day period it may have been that of his higher-dimensional being (his soul, his spiritual self) in a space of higher dimensions, a space radically different from our space of three dimensions. The impressions he received were therefore the impressions received in a space totally foreign to us, impressions which could not be described to us because we have no words, no language which will convey such a description to human (three-dimensional) beings. What is more, any such impressions that Lazarus may have been conscious of before Jesus raised him from the grave vanished the instant he came to life, that is, the instant he again became a mortal, a three-dimensional being. On being again clothed with his material body his spirit at once became subject to the restrictions and limitations imposed on beings living in our space of three dimensions.

It seems futile therefore to expect that we shall ever receive a message thru human agencies from those who have departed to the great beyond describing their state of existence because such a message can convey no meaning whatever to us. The Bible contains no promise of any such knowledge but on the contrary asserts that this must remain a sealed book to mortals. This does not, however, interpose any obstacles whatever to divine revelations to man, either in the past or in the future, because God is not limited in any manner by the restrictions of any space.

THE SACRAMENT OF THE LORD'S SUPPER

All the different shades of doctrine held by Christian believers regarding the sacrament of the Lord's Supper may be roughly divided into three distinct groups.

Group I. Doctrines which include transubstantiation as a fundamental principle. These assert that the bread and wine are literally transformed into the body and blood of Jesus. According to this principle of transubstantiation the believer partakes, not of the bread, but of the actual material human body of Jesus, and in drinking the cup drinks, not wine, but the actual material human blood of Jesus.

Group II. Doctrines which hold that the bread and wine only represent or symbolize the body and blood of Jesus. According to these the believer partakes only of the material bread and wine, the act being merely a remembrance, a celebration commemorating the last supper of Jesus with his disciples.

Group III. Doctrines which assert that the believer in partaking of the bread and wine also by the same act partakes not of the material body and blood of Jesus but of the spiritual body and blood of a resurrected, a glorified Jesus.

The first of these three groups of doctrines is naturally repellant to us because of the grossness of the act involved, namely, the consumption of human flesh and blood. It is the human body and blood of Jesus, into which the bread and wine has been transformed, of which the believer is supposed to partake.

On the other hand, the second of these three groups of doctrines does not entirely satisfy those who are spiritually minded because it minimizes the sacrament by making it merely a commemora tion ceremony. This viewpoint is difficult to reconcile with the words of Jesus, "this *is* my body this *is* my blood."

Defenders of the doctrine of transubstantiation invariably attempt to justify their belief in it by quoting the above words of Jesus. In so doing they are trying to support their case on the wholly unwarranted assumption that Jesus spoke of his material, his human body and blood. Jesus knew of his coming suffering, death and resurrection. He spoke of the future beyond Golgotha when he said "this do, as oft as ye drink it, in remembrance of me." In this future there would be no more a material, a human body and blood of Jesus, but only his spiritual, his glorified body and blood.

The third of the above groups of doctrines therefore is sanctioned by the words of Jesus. It also commends itself to human reason. It is therefore near at hand to assert that Jesus is present in the sacrament of the Lord's Supper as a higher-dimensional being. He descends from heaven (higher-dimensional space) to commune with the believer in our world (three-dimensional space). We cannot detect his presence thru our material senses because he is a higher-dimensional being while we are laboring under the restrictions and limitations imposed on three-dimensional beings. There is no necessity for the spacial presence (in the ordinary sense) of Jesus in order to fulfill his promise to be with us. How much greater and glorious as well as more satisfying to the believer is not this presence of the spiritual, the glorified (higher-dimensional) body and blood of Jesus than would be the presence of merely his human, his material body and blood!

CHAPTER XII

DESCRIPTION OF HEAVEN. MIRACLES

It is evident that we as three-dimensional beings cannot have a true conception of what heaven is; being a higher space it is beyond our powers of understanding. If an accurate description of heaven were in some way available we would not understand an iota of it, it would be meaningless to us "For since the beginning of the world men have not heard, nor perceived by the ear, neither hath the eye seen, O God, besides thee, what he hath prepared for him that waiteth for him" Isa. 64:4. "Eye hath not seen, nor ear heard, neither have entered into the heart of man, the things which God hath prepared for them that love Him." I Cor. 2:9. Christ spoke to his Apostles of the impossibility of their having any image or notion of the place to which when he disappeared he would go and whence he would return. Attempts to describe heaven always result in a description in terms of our three-dimensional environment. In this connection it is significant to note that the Bible usually tells what heaven is not rather than what heaven is. The adjectives are generally negative. We should never lose sight of the fact that the Bible was written for beings living in a three-dimensional space, beings whose knowledge and experiences are wholly limited to that space. Dr. P. Anstadt in his book "Recognition in Heaven" gives the following summary of Bible statements describing heaven·

Negative Features, or the Things that Will Not Be There	*Positive Features, or the Things that Will Be There.*
Indestructible.	The city of our God— the heavenly Jerusalem.
Undefilable.	
Unchangeable.	Beautiful waters.
No crying.	Delicious fruits.
No tears	Sure healing for the nations.
No pain.	
No sorrow.	Populous with happy people.
No death.	
No burning sun.	Beautiful garments.
No cold or heat.	Devout worship.
No night.	Enchanting music.
No hunger.	A just ruler.
No thirst.	An eternal kingdom.
No bad men.	The grandest capitol.
No sin.	Many mansions.
No curse.	

MIRACLES

One reason why some very good people hesitate about accepting the miracles of the Bible at their face value is that they assume that such miracles can only be brought about by violating or in some way interfering with the laws of nature, the laws governing our three-dimensional universe. This difficulty vanishes if instead we look upon the miracles of the Bible as the perfectly logical results of the working out of laws connected with higher spaces, laws which include the lower, the so-called natural laws of our material universe, just as our space is contained within the higher spaces. We may then consider the miracles performed by Jesus, the apostles and the prophets to consist of momentary revelations by God to mortals of the operations of some of those higher laws of which we are as yet ignorant. It is not at all probable that the apostles and prophets themselves understood these higher laws, and it was not necessary that they should, any more than that it is necessary for the operator who sends a message by wireless to understand the theory of wireless telegraphy or to have actual knowledge of the medium thru which it is transmitted. These higher laws may be the laws of four-dimensional or still higher spaces or they may be laws connected with realms of thought and being of which we now have no conception whatever. To insist that God can make wine out of water only by pouring fermented grape juice into it is ridiculous! If he were thus limited he would not be God. The rending of the veil of the Temple on the crucifixion of Christ and the speaking with many tongues on the giving out of the Holy Ghost, these may all be revelations to man of higher-dimensional phenomena.

Swedenborg's involved descriptions of "heavenly" forms, motions, and mechanics, become somewhat more intelligible when interpreted in terms of higher spaces even if we are not willing, as the author is not, to agree with his conclusions. Helmholtz continually kept the possibility of physical higher spaces before him in his dynamical studies, reasonings which are among the classics of physical science. Kelvin also felt the pressure of the reality of higher spaces so strongly that he declared himself ready to accept them as an explanation of physical phenomena when these could be more consistently explained by such a concept.

CHAPTER XIII

CONCLUSION

The new vistas of thought which have been opened to us by the higher space hypothesis have revealed unlimited realms of being and doing, all of which are interrelated in a manner far surpassing the powers of ingenuity of even a superman. In the light of this vision, how small and mean, how low and limited, yes, how contemptible, does not the materialist appear as he burrows into the dust in his attempts to discover the source of the glorious rays of truth and light which radiate from God! As we, in this study of higher spaces, have caught a momentary glimpse of the sublimity and vastness, the glory and grandeur, the symmetry and order, of what must be only a very small part, perhaps only an infinitesimal part, of creation; how preposterous does it not seem, yes, how ridiculous even to suppose that all should have come into being thru a succession of fortuitous circumstances, by blind chance, and not thru a supreme intelligence with no limitations of knowledge or power.

How pitiful and weak is man with his limitations of thought and action in the presence of these infinite possibilities! And yet there are men presumptuous enough to claim that human reason and understanding can account for it all. No matter how wise or powerful a man may be he is but one of billions of humans, to whom this our earth has been or is the dwelling place. And our earth is only one of the smallest of the planets revolving around our sun, every one of which may be inhabited by beings, not necessarily with bodies like ours, because they would have to be adapted to their own physical environment, but perhaps with souls like ours. Our sun, however, is only one of the smaller of the suns, fixed stars, apparently without number, which dot our heavens. In some parts of the sky there appear clusters of them, as in the Milky Way, so numerous that they appear only as patches of light. Some of these suns are so far away that it takes centuries for the light from them to reach us. A few years ago a star of very large magnitude suddenly appeared in our sky. It then grew gradually smaller until now it is not at all conspicuous. Astronomers told us that this new star really represented a conflagration in the heavens, two large suns had collided, but the collision had taken place over one hundred years before, at the very time when Napoleon was walking the world like a Colossus, and the flash of light caused by the impact never reached earth until in our day. In fact, there may be suns so far from us and created so long ago that light from them has not yet reached the earth. Each of these countless suns may have many more and larger planets revolving about them than is the case with our sun, and each planet may be inhabited by beings with souls. All of this vast infinitude that we have been contemplating exists, however, simply in our space of three dimensions. There may exist an infinite number of other three-dimensional spaces, each one comprising a universe as great as or greater than our own. But enfolding these three-dimensional spaces as in a garment may be a four-dimensional space, and that may be only one of an infinite number of other possible four-dimensional spaces. Again, these four-dimensional spaces may be enfolded by five-dimensional spaces; and so on thru spaces of six, seven, etc., dimensions up to space of infinite dimensions, where God himself dwells subject to no space limitations.

But now man, at the thought of these possibilities, stands aghast, stupified; he exclaims, "it can't be true, it is all too vast, too stupendous to lie within the realms of the possible." Even some who believe in God are inclined to think that this concept exceeds the bounds of the attainable. But by thus placing a limit to what God can do we are virtually denying that he is God. Instead of being restricted to those possibilities which may be humanly conceived, God is able to put our unbelief to shame by exhibitions of power of which we have now not the slightest conception.

INDEX

How to update your TV APP

1 ~~M~~ MAKE SURE you give STORAGE Permission

2 PI

_____

SETTING → DEVICE Develop options

APPS From unknow Sources

Made in the USA
Columbia, SC
30 December 2017